# Basics about Robotics

# Anish K Raj

Recognizing God's literary grace

I am grateful to my mother's, K G Kavitha, Dr. K G Anitha and K G Nanitha whose unfailing encouragement and support sustained me through the challenge of writing this book. There are no words that can express my heartfelt gratitude to members of C V Raman Global University, in particular, Dr. Ranjitha Mohanty, Dr. Bikash Ranjan Moharana and Dr. Kalinga Simant Bal, who encouraged me enormously during the realization of this book.

# Contents

# Roadmap

In recent years, robotics has emerged as a transformative technology with the potential to revolutionize various industries. From manufacturing and healthcare to space exploration and everyday life, robots are increasingly being integrated into our world, performing tasks with precision, speed, and efficiency. The book "Advances in Robotics: Exploring the Future of Automation" dives deep into the realm of robotics, offering an in-depth exploration of the current advancements and future possibilities in this rapidly evolving field.

This comprehensive book is a culmination of the latest research and developments in robotics, presenting a holistic view of the subject matter. It is intended to serve as a valuable resource for students, researchers, and professionals seeking to expand their knowledge and understanding of robotics.

Chapter by chapter, "Advances in Robotics" delves into various aspects of this multifaceted discipline, examining the intricate components, mechanisms, and functionalities that enable robots to perceive, manipulate, and interact with their environment. The book also delves into the underlying theories and algorithms that drive robot control, programming, and planning.

The introductory chapter sets the stage by providing a solid foundation in robotics. It offers a historical overview of the field, tracing its roots and highlighting key milestones that have shaped its trajectory. The chapter also explores the fundamental principles and concepts that underpin robotics, including the integration of mechanical, electrical, and computer engineering.

Moving forward, subsequent chapters dissect different facets of robotics, providing detailed insights into robot components, locomotion, manipulation, perception, control, programming, kinematics, sensors, vision, communication, and planning. Each chapter is written by experts in their respective fields, ensuring the highest level of expertise and accuracy in the information presented.

The book adopts an academic style, adhering to rigorous scientific standards. It combines theoretical discussions with practical applications, striking a balance between conceptual understanding and real-world implementation. Mathematical formulations, diagrams, and illustrations are utilized to clarify complex concepts and facilitate comprehension.

By examining the current state-of-the-art in robotics, "Advances in Robotics" aims to inspire readers to envision the future potential of automation. The book not only explores the existing capabilities of robots, but also probes the frontiers of what is yet to come. It delves into emerging technologies, such as artificial intelligence, machine learning, and human-robot interaction, that are poised to redefine the boundaries of robotics.

In conclusion, "Advances in Robotics: Exploring the Future of Automation" presents a comprehensive and authoritative exploration of robotics. Through its meticulously crafted chapters, this book invites readers to embark on an enlightening journey into the world of robots, unveiling the intricacies of their design, function, and application. Whether you are a student, researcher, or professional, this book will serve as an invaluable resource, equipping you with the knowledge and insights to navigate the exciting landscape of robotics.

# Chapter 1

# Introduction to Robotics

## 1.1 The Evolution of Robotics

The field of robotics has witnessed a remarkable evolution over the years, transforming from simple mechanical contraptions to sophisticated and intelligent machines. The roots of robotics can be traced back to ancient times, where inventors and engineers experimented with automata, mechanical devices capable of performing predetermined actions. However, it was not until the 20th century that significant advancements in technology paved the way for the birth of modern robotics.

## 1.2 The Interdisciplinary Nature of Robotics

Robotics is inherently interdisciplinary, drawing upon various fields of study such as mechanical engineering, electrical engineering, computer science, and artificial intelligence. This multidisciplinary approach allows for the integration of diverse expertise and the development of comprehensive robotic systems.

## 1.3   The Components of a Robot

A robot consists of several key components that work together
to enable its functionality. These components typically include
a mechanical structure, sensors, actuators, power supply, and a
control system. The mechanical structure provides the physical framework of the robot, allowing it to interact with its
surroundings. Sensors act as the robot's sensory organs, gathering data about the environment. Actuators, such as motors
and servos, enable the robot to move and manipulate objects.
The power supply provides the necessary energy to drive the
robot's components, while the control system governs its behavior and actions.

## 1.4   Robot Locomotion

One of the fundamental aspects of robotics is locomotion,
which refers to the ability of a robot to move and navigate
its environment. Robots employ various locomotion mechanisms depending on their design and intended application.
These mechanisms can range from wheels and tracks to legs
and even flying systems. Each method of locomotion presents
its own advantages and challenges, and researchers continue
to explore innovative approaches to enhance robot mobility.

## 1.5   Robot Manipulation

Robot manipulation entails the ability of a robot to interact
with and manipulate objects in its environment. This involves

tasks such as grasping, picking, placing, and manipulating objects with precision and dexterity. Advances in robotic manipulation have been instrumental in revolutionizing industries such as manufacturing, where robots now play a significant role in assembly lines and repetitive tasks.

## 1.6   Robot Perception

Perception is a critical aspect of robotics that enables robots to understand and interpret their surroundings. Through the use of sensors such as cameras, LIDAR, and proximity sensors, robots can gather data about the environment, including object detection, recognition, and localization. Perception algorithms and techniques continue to advance, enabling robots to perceive the world with increasing accuracy and efficiency.

## 1.7   Robot Control

Robot control involves the regulation and coordination of a robot's actions and movements. It encompasses both low-level control, which focuses on individual components and actuators, and high-level control, which governs the overall behavior and decision-making of the robot. Control systems can range from simple rule-based algorithms to complex artificial intelligence algorithms that enable robots to adapt to changing circumstances and make autonomous decisions.

## 1.8   Robot Programming

Programming robots involves specifying the instructions and commands that dictate their behavior and actions. Robot programming languages provide a means to communicate with robots and define their tasks. These languages can range from traditional programming languages like C++ and Python to domain-specific languages tailored for robotics. The development of intuitive and user-friendly programming interfaces has played a crucial role in democratizing robotics and making it accessible to a broader audience.

## 1.9   Robot Kinematics

Kinematics deals with the study of the motion of robots, focusing on the relationships between their components, joints, and end effectors. Kinematic analysis enables engineers to model and simulate the movement of robots, aiding in the design and optimization of their mechanical structures. Kinematic algorithms are essential for tasks such as trajectory planning, inverse kinematics, and forward kinematics, which determine the position and orientation of a robot's end effector based on the joint angles.

## 1.10   Robot Sensors

Sensors play a vital role in robotics by providing robots with the ability to perceive and interact with their environment. A wide range of sensors is employed in robotics, including cameras, range finders, force sensors, and tactile sensors. These sensors enable robots to gather data about the surrounding objects, detect obstacles, measure distances, and even sense temperature and humidity. Sensor fusion techniques, which integrate data from multiple sensors, enhance the robot's perception capabilities and enable robust decision-making.

# 1.11　Robot Vision

Robot vision refers to the ability of robots to perceive and understand visual information from their surroundings. By utilizing cameras and image processing algorithms, robots can analyze images and extract valuable data for object recognition, tracking, and scene understanding. Vision-based systems have found applications in various domains, including autonomous navigation, object detection, and industrial quality control.

# 1.12　Robot Communication

Communication is crucial for robots to collaborate with humans and other robots effectively. Robot communication involves the exchange of information between robots and their operators or between multiple robots within a system. Communication protocols, such as Ethernet, Wi-Fi, and Bluetooth, facilitate seamless data transfer and enable coordination and cooperation among robots. Additionally, advances in natural language processing and human-robot interaction have facilitated more intuitive and efficient communication between humans and robots.

# 1.13　Robot Planning

Robot planning involves the generation of optimal paths and sequences of actions to accomplish specific tasks or goals. Planning algorithms consider various factors, including the robot's capabilities, the environment, and any constraints or obstacles present. Path planning, motion planning, and task planning are essential components of robot planning, enabling robots to navigate complex environments, avoid collisions, and execute tasks efficiently.

## 1.14 Robot Applications

The application domains of robotics are diverse and continue to expand rapidly. Robots have been deployed in industries such as manufacturing, healthcare, agriculture, logistics, space exploration, and entertainment. In manufacturing, robots have revolutionized production processes, increasing efficiency, accuracy, and productivity. In healthcare, robots assist in surgery, rehabilitation, and patient care. They can also be found in agriculture, where they automate tasks such as harvesting and monitoring crop health. Furthermore, robots are increasingly used in exploration missions to other planets and moons, enabling scientists to gather data in extreme environments.

In conclusion, robotics represents a fascinating and rapidly evolving field with vast potential for innovation and societal impact. "Advances in Robotics: Exploring the Future of Automation" aims to provide readers with a comprehensive understanding of the various aspects of robotics, from its historical roots to the latest advancements in robot components, locomotion, manipulation, perception, control, programming, kinematics, sensors, vision, communication, and planning. By delving into the intricacies of robotics, this book aims to inspire and inform researchers, students, and professionals in their pursuit of pushing the boundaries of automation and shaping the future of robotics.Mair 1988; Vertut and Coiffet 1986; Niku 2020

# Chapter 2

# Robot Components

## 2.1 Introduction to Robot Components

Robots consist of various components that work together to enable their functionality. These components can be categorized into mechanical, electrical, and computational elements. Understanding the different components is crucial for designing, building, and operating robots effectively.

## 2.2 Mechanical Components

Mechanical components form the physical structure of a robot. They provide the framework for supporting other components and enable the robot to interact with its environment. Common mechanical components include frames, joints, linkages, and end effectors.

The frame of a robot serves as its skeleton, providing rigidity and structural support. Frames are typically constructed from materials such as aluminum, steel, or composite materials, balancing strength and weight considerations. The frame design depends on the intended application of the robot, whether it is a humanoid robot, a robotic arm, or a mobile robot.

Joints connect the different parts of a robot's body, allowing relative motion between them. Depending on the type of joint, robots can have rotational joints (e.g., revolute joints for arm rotation) or translational joints (e.g., prismatic joints for linear motion). The number and configuration of joints determine the robot's degrees of freedom (DOF) and its range of motion.

Linkages are the rigid or flexible elements that connect the joints and transfer motion. They can take the form of arms, limbs, or other mechanical structures, and their design depends on the desired motion and dexterity of the robot. Linkages may incorporate gears, belts, or pulleys to transmit motion and amplify forces.

End effectors, also known as robot hands or grippers, are the components responsible for interacting with objects in the environment. End effectors come in various forms, such as robotic fingers, suction cups, or specialized tools. The design of the end effector depends on the tasks the robot needs to perform, whether it is grasping, manipulating, or performing delicate operations.

## 2.3 Electrical Components

Electrical components in robots are responsible for power distribution, control signals, and data transmission. These components include batteries or power supplies, motors, sensors, and actuators.

Batteries or power supplies provide the necessary electrical energy to power the robot's components. The selection of batteries depends on factors such as the robot's power requirements, operating duration, and portability considerations. Advanced power management systems are employed to optimize energy usage and extend the robot's operating time.

Motors are essential components for generating mechanical motion in robots. DC motors, stepper motors, and servo motors are commonly used in robotics. DC motors provide continuous rotation, while stepper motors offer precise control over angular displacement. Servo motors combine the advantages of both, providing accurate positioning and torque control.

Sensors are crucial for robots to perceive the environment and gather data. Robots can be equipped with a wide array of sensors, including proximity sensors, force sensors, temperature sensors, and inertial sensors. Proximity sensors detect the presence of objects, while force sensors measure applied forces or tactile feedback. Temperature sensors monitor temperature variations in the robot or its surroundings. Inertial sensors, such as accelerometers and gyroscopes, provide information about the robot's motion, orientation, and acceleration.

Actuators are responsible for converting electrical energy into mechanical motion. They include motors, pneumatic actuators, hydraulic actuators, and shape-memory alloys. Pneumatic and hydraulic actuators provide high forces and are suitable for applications requiring significant strength. Shape-memory alloys can change their shape when heated, offering unique actuation capabilities.

## 2.4   Computational Components

Computational components in robots handle the processing and control aspects. These components include microcontrollers, microprocessors, onboard computers, and software systems.

Microcontrollers and microprocessors serve as the "brains" of a robot, responsible for executing control algorithms and coordinating the robot's actions. They process sensor data, implement control strategies, and generate output signals to actuate the robot's components. These computational components are designed to be compact, low-power, and capable of real-time processing.

Onboard computers are more powerful computational units that enable complex computations and higher-level decision-making. They provide the computational resources for running advanced algorithms, such as machine learning or computer vision, and facilitate more sophisticated robotic behaviors. Onboard computers often feature multiple processors, high-speed interfaces, and sufficient memory capacity to handle the computational demands of modern robotics applications.

Software systems play a crucial role in robot control and operation. These systems encompass operating systems, middleware, and application software. Operating systems provide an interface between the hardware components and the software applications, managing resources, scheduling tasks, and facilitating communication. Middleware offers a set of software libraries and tools for developing and integrating robot applications. Application software encompasses the specific algorithms, control strategies, and user interfaces tailored to the robot's intended tasks and functionalities.

## 2.5   Integration of Components

The successful integration of the various robot components is essential for achieving a cohesive and functional robotic system. Mechanical components must be carefully designed and assembled to ensure proper fit, stability, and range of motion.

Electrical components need to be properly connected and powered, with wiring and circuitry organized to minimize interference and maximize reliability. Computational components require efficient programming, data processing, and control algorithms to achieve the desired robot behaviors.

The integration process involves designing interfaces and connectors that enable seamless communication and interoperability between components. It also entails the calibration and alignment of sensors and actuators to ensure accurate data acquisition and precise motion control. Comprehensive testing and debugging are conducted to identify and resolve any integration issues or malfunctions.

Furthermore, the integration process should consider factors such as weight distribution, power consumption, and thermal management to ensure the overall performance and safety of the robot. Robust mechanical and electrical designs, as well as effective software architecture, are essential for achieving a well-integrated and reliable robotic system.

## 2.6    Advances in Robot Component Technologies

Robot component technologies are continuously advancing, driven by ongoing research and development efforts. Innovations in materials science have led to the development of lightweight and durable materials, allowing for the construction of more agile and energy-efficient robots. New manufacturing techniques, such as additive manufacturing (3D printing), enable the rapid prototyping and customization of robot components.

In the field of electrical components, advancements in battery technology have led to the development of more efficient and high-capacity power sources, enabling longer operating times for robots. Miniaturization of sensors and improvements in sensor technology have resulted in more compact and accurate sensing capabilities. Additionally, advancements in actuator technology, such as the development of soft and flexible actuators, offer new possibilities for compliant and safe human-robot interaction.

Computational components continue to evolve with the advancement of microprocessor technology and the integration of artificial intelligence and machine learning algorithms. These developments enable robots to perform complex tasks, adapt to dynamic environments, and learn from experience. Furthermore, the emergence of cloud computing and edge computing paradigms allows for distributed computing and resource sharing, enhancing the computational capabilities of robots.

In conclusion, the components of a robot form the foundation for its functionality and capabilities. Mechanical components provide the physical structure and allow for interaction with the environment, while electrical components enable power distribution and signal processing. Computational components serve as the brain of the robot, executing control algorithms and facilitating higher-level decision-making. The successful integration of these components is crucial for the development of functional and efficient robotic systems. With advancements in materials science, electrical technology, and computational power, the field of robotics continues to witness remarkable advancements in robot component technologies. These advancements have led to the development of more capable, versatile, and intelligent robots across various industries and applications.

In recent years, there have been significant breakthroughs in materials science that have revolutionized robot construction. Lightweight and strong materials, such as carbon fiber composites, have allowed for the creation of robots that are both robust and agile. These materials offer improved strength-to-weight ratios, enabling robots to carry heavier payloads while maintaining maneuverability. Additionally, the use of flexible and stretchable materials has led to the development of soft robots, which possess inherent compliance and can safely interact with humans in sensitive environments.

Advancements in electrical components have greatly enhanced the power and efficiency of robots. Lithium-ion batteries, for instance, provide higher energy density and longer operational times, enabling robots to operate for extended periods without frequent recharging. This has proven particularly beneficial in applications where robots need to perform tasks autonomously or in remote locations. Furthermore, improvements in motor technology, such as brushless DC motors and high-torque servos, have resulted in increased precision, speed, and torque control, allowing robots to perform delicate and complex movements with greater accuracy.

Sensors play a critical role in enabling robots to perceive and interact with their environment. Recent developments in sensor technology have expanded the sensing capabilities of robots, enabling them to gather more detailed and accurate data. For instance, advancements in computer vision have led to the development of high-resolution cameras and depth sensors that can accurately perceive objects and environments in three dimensions. Infrared sensors and thermal imaging technology have enhanced robots' ability to detect and respond to changes in temperature, making them suitable for applications such as industrial inspections or search and rescue operations.

The integration of artificial intelligence (AI) and machine learning algorithms has revolutionized the computational capabilities of robots. AI-powered robots can learn from data, adapt to changing conditions, and make intelligent decisions. Machine learning algorithms enable robots to recognize patterns, classify objects, and improve their performance over time. Reinforcement learning techniques have allowed robots to autonomously learn and refine their behaviors through trial and error, enabling them to navigate complex environments or manipulate objects with greater proficiency.

Furthermore, advancements in cloud computing and edge computing have opened up new possibilities for robot capabilities. Cloud computing allows robots to leverage the computational power and storage resources of remote servers, enabling them to perform complex computations and access vast amounts of data. Edge computing, on the other hand, enables robots to process data locally, reducing latency and enabling real-time decision-making. These computing paradigms have facilitated the development of collaborative robots that can offload computationally intensive tasks or share knowledge with other robots in a networked environment.

In conclusion, the field of robotics continues to witness remarkable advancements in robot component technologies. These advancements in materials science, electrical components, sensor technology, and computational power have led to the development of more capable, versatile, and intelligent robots. As technology continues to evolve, robots are becoming increasingly integrated into our daily lives, transforming industries, enhancing productivity, and revolutionizing the way we interact with machines. The ongoing progress in robot component technologies holds great potential for the future of robotics, where robots will continue to play a significant role in addressing complex challenges and improving the quality of our lives.Katevas 2001; Nehmzow 2012; Corke 2021

# Chapter 3

# Robot Locomotion

## 3.1 Introduction to Robot Locomotion

Robot locomotion refers to the ability of a robot to move from one place to another in its environment. Locomotion is a fundamental aspect of robotics and plays a crucial role in enabling robots to perform tasks and interact with the world around them. This chapter explores various methods and technologies used in robot locomotion, ranging from wheeled and legged locomotion to aerial and underwater locomotion.

## 3.2 Wheeled Locomotion

Wheeled locomotion is one of the most common and straightforward methods employed by robots. It involves the use of wheels or tracks to propel the robot and navigate across different surfaces. Wheeled robots are highly versatile and efficient in environments with flat and smooth terrains.

Wheeled robots can be further categorized into differential

drive, omnidirectional, and holonomic drive systems. Differential drive systems utilize two or more wheels, with each wheel driven independently, allowing the robot to perform turns by varying the speed and direction of the wheels on each side. Omnidirectional drive systems incorporate wheels that can move independently in any direction, enabling the robot to move laterally and perform complex maneuvers. Holonomic drive systems use wheels or rollers arranged in a specific configuration to achieve full maneuverability in any direction, including translations and rotations.

## 3.3    Legged Locomotion

Legged locomotion aims to mimic the walking or running motion of animals, offering robots the ability to navigate various terrains, including uneven or rough surfaces. Legged robots possess the advantage of traversing obstacles and adapting to challenging environments that are inaccessible to wheeled robots. The complexity and versatility of legged locomotion make it a highly active research area in robotics.

Legged robots can have different leg configurations, such as bipedal (two legs), quadrupedal (four legs), or hexapodal (six legs). Each leg is typically equipped with joints and actuators to control its motion. The control of legged locomotion involves coordination and synchronization of the leg movements, balancing, and adaptive gait generation.

Researchers have developed various approaches to legged locomotion, including dynamic walking, passive dynamic walking, and compliant locomotion. Dynamic walking relies on actively controlling the robot's movements, applying forces and torques to achieve stability and efficient locomotion. Passive dynamic walking, on the other hand, utilizes the natural dynamics of the robot's body and gravity to generate walking motion without active control. Compliant locomotion focuses on utilizing compliant elements and elasticity in the legs to absorb impacts and improve energy efficiency during locomotion.

## 3.4   Aerial Locomotion

Aerial locomotion involves robots that are capable of flying or hovering in the air. This mode of locomotion enables robots to access elevated areas, survey large areas quickly, and overcome obstacles on the ground. Aerial robots, commonly known as drones or unmanned aerial vehicles (UAVs), have gained significant popularity and have become increasingly accessible for various applications.

Aerial robots can have different configurations, such as fixed-wing, rotary-wing, or flapping-wing. Fixed-wing drones resemble miniature airplanes and rely on aerodynamic lift generated by their wings to maintain flight. Rotary-wing drones, often referred to as quadcopters, utilize multiple rotors to generate lift and control their movement in the air. Flapping-wing robots mimic the flight of birds or insects, utilizing the flapping motion to generate lift and propulsion.

Aerial locomotion poses unique challenges in terms of stability, control, and energy efficiency. Maintaining stability in the air requires precise control of the drone's attitude (orientation) and altitude. Control systems based on inertial measurement units (IMUs), GPS, and onboard sensors enable stable flight and accurate navigation.

Aerial robots must also consider energy efficiency to maximize flight time. The design of lightweight structures, efficient propulsion systems, and optimized control algorithms contribute to longer endurance and improved maneuverability. Additionally, advancements in battery technology have led to the development of high-capacity and lightweight power sources, enabling drones to operate for extended periods.

Applications of aerial locomotion are diverse, ranging from aerial surveillance and photography to delivery services and search and rescue operations. Drones equipped with cameras and sensors can capture aerial images and videos for mapping, inspection, and monitoring purposes. They can be deployed in disaster-stricken areas to assess damages and locate survivors. In the field of agriculture, aerial robots can aid in crop monitoring, pest control, and precision spraying. Aerial locomotion also offers new perspectives in the entertainment industry, with the use of drones for aerial shows and light displays.

## 3.5   Underwater Locomotion

Underwater locomotion involves robots designed to operate in aquatic environments, such as oceans, lakes, or underwater structures. These robots, often referred to as autonomous underwater vehicles (AUVs) or underwater drones, are equipped with specialized features that allow them to navigate, explore, and collect data in underwater environments.

AUVs can be categorized into different types based on their propulsion systems and locomotion methods. Propeller-driven AUVs utilize one or more propellers to generate thrust for propulsion, allowing them to move through the water in a controlled manner. Flapping fin AUVs imitate the movements of marine animals, utilizing flexible fin-like structures to generate forward propulsion. Biomimetic AUVs, inspired by the locomotion of marine creatures, mimic the motion of fish, octopuses, or other aquatic organisms to achieve efficient and agile underwater locomotion.

Underwater locomotion presents unique challenges due to the harsh conditions and constraints imposed by the underwater environment. These challenges include buoyancy control, pressure resistance, and communication limitations. AUVs employ various methods to achieve buoyancy control, such as ballast systems that allow them to adjust their overall weight and maintain neutral buoyancy. Specialized materials and design considerations are required to ensure the structural integrity of the robot under high-pressure conditions at greater depths. Additionally, communication with underwater robots often relies on acoustic signals due to the limited range and signal propagation properties underwater.

The applications of underwater locomotion are wide-ranging. AUVs are used for marine research, oceanographic studies, and environmental monitoring. They can collect data on water quality, marine life, and underwater geological features. AUVs equipped with imaging systems can capture high-resolution images and videos of underwater environments for scientific research or underwater archaeology. They are also employed in underwater inspections of infrastructure, pipelines, and submerged structures.

In conclusion, robot locomotion encompasses various methods and technologies that enable robots to move and navigate in their respective environments. Whether through wheeled locomotion for flat terrains, legged locomotion for traversing challenging surfaces, aerial locomotion for accessing elevated areas, or underwater locomotion for exploring aquatic environments, each mode of locomotion presents unique opportunities and challenges. Advances in locomotion technologies continue to push the boundaries of what robots can achieve, opening up new possibilities for applications in fields such as exploration, surveillance, agriculture, and beyond. The continuous development of locomotion systems will contribute to the further advancement and integration of robots into our daily lives.Murphy 2019; Kelly 2013; Jazar 2010

# Chapter 4

# Robot Manipulation

## 4.1 Introduction to Robot Manipulation

Robot manipulation refers to the ability of robots to interact with and manipulate objects in their environment. It involves a wide range of tasks, including grasping, picking, placing, and manipulating objects with precision and dexterity. The field of robot manipulation is essential for various industries, including manufacturing, logistics, healthcare, and household robotics.

## 4.2 Gripper Systems

Gripper systems are an integral part of robot manipulation, as they provide the means to grasp and hold objects securely. Grippers come in different types, each suited for specific applications and object characteristics. Some common gripper types include:

1. **Parallel-jaw Grippers:** These grippers feature two parallel jaws that move towards each other to grasp objects. They are versatile and suitable for a wide range of object shapes and sizes. Parallel-jaw grippers can be designed with various gripping mechanisms, such as pneumatic, hydraulic, or electric.

2. **Vacuum Grippers:** Vacuum grippers use suction to hold objects. They are particularly effective for flat and smooth objects, such as sheets of paper or glass panels. Vacuum grippers typically employ a vacuum pump to create suction and special suction cups or pads to adhere to the object's surface.

3. **Mechanical Grippers:** Mechanical grippers utilize mechanical mechanisms, such as fingers or claws, to grip objects. They often provide a strong and secure grasp, making them suitable for handling heavy or irregularly shaped objects. Mechanical grippers can be actuated using pneumatic, hydraulic, or electric systems.

## 4.3   Perception for Manipulation

Perception plays a crucial role in robot manipulation by providing the necessary information about objects' positions, shapes, and properties. Perception systems enable robots to identify and locate objects, assess their properties (such as size, shape, and material), and plan appropriate manipulation strategies.

Vision-based perception is commonly used in robot manipulation tasks. Cameras mounted on robots capture images or video streams, which are then processed using computer vision algorithms. These algorithms analyze the images to detect and recognize objects, estimate their poses, and extract relevant features. Depth sensors, such as stereo cameras or time-of-flight cameras, provide additional depth information, enabling robots to perceive objects in three-dimensional space.

Other sensing modalities, such as force/torque sensors and tactile sensors, can provide valuable feedback during manipulation tasks. Force/torque sensors measure the forces and torques applied during object manipulation, allowing robots to adapt their grip and exert appropriate forces. Tactile sensors, embedded in grippers or robot fingers, provide information about contact forces and object properties, enabling delicate and precise manipulation.

## 4.4 Manipulation Planning and Control

Manipulation planning involves generating a sequence of actions that enable a robot to achieve a desired manipulation task. It encompasses tasks such as reaching a target pose, grasping an object, and manipulating it to a desired location or orientation. Planning algorithms consider the robot's kinematics, the object's geometry, and any constraints or obstacles in the environment.

Grasping planning focuses on finding appropriate grasps that allow the robot to securely hold an object. It takes into account the object's shape, size, and stability, as well as the robot's kinematic capabilities. Grasp planning algorithms aim to optimize the stability and manipulability of the grasp, considering factors such as contact forces, friction, and object stability.

Once a grasp is established, manipulation control comes into play. Control algorithms determine the robot's movements and actuator commands to execute the desired manipulation actions. These algorithms consider factors such as trajectory planning, collision avoidance, force control, and compliance control. They ensure that the robot performs manipulation tasks accurately, safely, and with the desired level of force and precision.

# 4.5 Advances in Robot Manipulation

Robot manipulation has witnessed significant advancements in recent years, driven by advancements in perception, control, and machine learning techniques. These advancements have enabled robots to perform more complex and delicate manipulation tasks with improved accuracy and efficiency. Here are some notable advances in robot manipulation:

1. **Learning-based Grasping:** Traditional grasp planning algorithms rely on explicit models of objects and environments. However, learning-based approaches have emerged, where robots learn to grasp objects through trial and error or with the help of simulated training. Machine learning techniques, such as deep neural networks, have been employed to learn grasp configurations that result in stable and successful grasps. These approaches have shown promising results in handling novel objects and adapting to different object shapes and sizes.

2. **Soft Robotics:** Soft robotics is an emerging field that focuses on the development of robots with soft and flexible structures and actuators. Soft robots have the advantage of being more adaptable and compliant, allowing for safe and delicate object manipulation. Soft grippers, made of flexible materials or with inflatable chambers, can conform to object shapes and provide gentle grasping capabilities. The integration of soft robotics with traditional rigid robots opens up new possibilities for

versatile and human-friendly manipulation.

3. **Cooperative Manipulation:** Cooperative manipulation involves multiple robots working together to manipulate objects that are too heavy or large for a single robot to handle alone. By coordinating their actions, robots can achieve complex manipulation tasks more efficiently. Cooperative manipulation requires careful coordination of grasping, lifting, and object manipulation among the robots. It finds applications in scenarios such as warehouse automation, construction, and assembly lines.

4. **Skill Transfer and Generalization:** Skill transfer and generalization techniques aim to enable robots to apply learned manipulation skills to new objects or tasks. Instead of learning from scratch for each new object or task, robots can leverage previously acquired knowledge and adapt it to similar scenarios. This involves capturing and representing manipulation skills in a generalized form, allowing robots to transfer and apply them to different objects or environments. Skill transfer and generalization techniques contribute to faster learning and more efficient manipulation in real-world settings.

5. **Dexterous Manipulation:** Dexterous manipulation refers to the ability of robots to manipulate objects with fine-grained control and precision, akin to human dexterity. Advances in sensor technology, control algorithms, and tactile feedback enable robots to perform delicate manipulation tasks such as precise object repositioning, manipulation of small or fragile objects, or assembly of intricate components. Dexterous manipulation finds applications in industries such as electronics assembly, healthcare, and scientific research.

6. **Human-Robot Collaboration:** Human-robot collab-

oration is an area of research that focuses on enabling robots to work alongside humans in shared workspaces. In collaborative manipulation, robots interact with human operators to perform tasks that combine human expertise with robotic precision and strength. This involves intuitive interfaces, safe interaction mechanisms, and shared autonomy to ensure effective collaboration. Human-robot collaboration has applications in industries such as manufacturing, healthcare, and rehabilitation.

These advances in robot manipulation have opened up new possibilities for automation, productivity enhancement, and improved human-robot interaction. As technology continues to progress, we can expect further advancements in robot manipulation, leading to more capable, adaptable, and versatile robots in the future. These robots will play a crucial role in various domains, contributing to increased efficiency, safety, and convenience in our daily lives.Bräunl 2003; Selig 2005; Unbehauen 2009

# Chapter 5

# Robot Perception

## 5.1 Introduction to Robot Perception

Robot perception refers to the ability of robots to perceive and understand their environment using sensors and algorithms. Perception plays a vital role in enabling robots to interact with and navigate through the world around them. By acquiring and interpreting sensory information, robots can make informed decisions, adapt to changing circumstances, and perform complex tasks.

## 5.2 Sensor Technologies for Robot Perception

Various sensor technologies are employed in robot perception, each providing unique types of information about the environment. Some commonly used sensor types include:

1. **Vision Sensors:** Vision sensors, such as cameras, are

widely used in robot perception. They capture images or video streams, which are then processed using computer vision algorithms. Vision sensors provide visual information about the surrounding objects, their positions, shapes, colors, and textures. They enable robots to recognize objects, track their movements, and navigate through complex environments.

2. **Range Sensors:** Range sensors, such as LiDAR (Light Detection and Ranging) and depth cameras, measure the distance between the robot and objects in its surroundings. LiDAR sensors emit laser beams and measure the time it takes for the beams to bounce back, allowing for the creation of detailed 3D maps of the environment. Depth cameras use infrared light or structured light patterns to estimate depth information, enabling the perception of objects in three dimensions.

3. **Tactile Sensors:** Tactile sensors provide robots with a sense of touch, allowing them to perceive and respond to physical contact with objects. These sensors can be in the form of sensitive skins, force sensors, or pressure-sensitive materials. Tactile information helps robots in tasks such as object manipulation, grasping, and interaction with humans or delicate objects.

4. **Proximity Sensors:** Proximity sensors, such as ultrasonic or infrared sensors, detect the presence or proximity of objects without physical contact. These sensors emit signals and measure the time it takes for the signals to bounce back, providing information about the distance between the robot and nearby objects. Proximity sensors are commonly used for obstacle detection and collision avoidance.

5. **Inertial Sensors:** Inertial sensors, such as accelerom-

eters and gyroscopes, measure the robot's acceleration, orientation, and angular velocity. They provide information about the robot's motion and its changes in position and orientation. Inertial sensors are often used in combination with other sensors to improve localization, mapping, and navigation capabilities.

# 5.3 Perception Algorithms and Techniques

Perception algorithms and techniques process the sensor data to extract meaningful information about the environment. These algorithms involve a combination of signal processing, pattern recognition, machine learning, and statistical techniques. Some commonly used perception techniques include:

1. **Object Recognition:** Object recognition algorithms analyze visual data to identify and categorize objects in the robot's field of view. These algorithms can utilize features such as shape, color, texture, and context to recognize objects and assign them to specific categories or classes. Object recognition is essential for tasks such as object manipulation, scene understanding, and navigation.

2. **Simultaneous Localization and Mapping (SLAM):** SLAM algorithms enable robots to build maps of their environment while simultaneously estimating their own position within the map. By integrating sensor data, such as range measurements or visual data, SLAM algorithms can create accurate representations of the environment and localize the robot within it. SLAM is crucial for autonomous navigation, exploration, and path planning.

3. **Point Cloud Processing:** Point cloud processing techniques are used to analyze data from range sensors, such as LiDAR or depth cameras. These algorithms process the 3D point cloud data to extract features, segment objects, and estimate object poses and shapes. Point cloud processing is widely employed in robotics applications such as object recognition, mapping, and autonomous driving.

4. **Semantic Segmentation:** Semantic segmentation algorithms assign semantic labels to individual pixels or regions in an image. By understanding the scene's semantic meaning, robots can differentiate between different objects, identify regions of interest, and make more informed decisions. Semantic segmentation is crucial for tasks such as scene understanding, object detection, and navigation planning.

5. **Object Tracking:** Object tracking algorithms enable robots to track the movement of specific objects over time. By associating objects across consecutive frames, these algorithms can estimate object trajectories, predict future positions, and maintain a consistent understanding of the environment. Object tracking is essential for tasks such as object manipulation, surveillance, and human-robot interaction.

6. **Sensor Fusion:** Sensor fusion techniques integrate data from multiple sensors to create a unified perception of the environment. By combining information from vision sensors, range sensors, and other modalities, robots can overcome the limitations of individual sensors and obtain a more comprehensive understanding of the surroundings. Sensor fusion enables tasks such as robust object detection, accurate localization, and reliable mapping.

# 5.4 Challenges and Future Directions

While significant progress has been made in robot perception, several challenges and opportunities for improvement remain. Some of the key challenges in robot perception include:

1. **Real-time Processing:** Perception algorithms often need to operate in real-time to enable timely and responsive robot actions. Efficient algorithms and hardware acceleration techniques are necessary to handle the large amount of sensor data and perform complex computations within tight time constraints.

2. **Robustness to Variability:** Environmental conditions, lighting changes, occlusions, and variations in object appearances pose challenges to perception systems. Developing robust algorithms that can handle such variability and generalize well across different environments and objects is crucial.

3. **Uncertainty and Ambiguity:** Perception is inherently uncertain, and sensor data can contain noise, outliers, and ambiguities. Developing probabilistic models and techniques to handle uncertainty, fuse information from multiple sources, and make reliable decisions is an ongoing research area.

4. **Lifelong Learning:** Enabling robots to learn and adapt their perception capabilities over time is essential. Lifelong learning techniques that allow robots to incrementally acquire new knowledge, update models, and adapt to changes in the environment can enhance their perception and interaction abilities.

5. **Multimodal Perception:** Integrating information from multiple sensory modalities, such as vision, touch, and

sound, can lead to richer and more robust perception. Developing algorithms and frameworks that can effectively fuse and exploit multimodal data is a promising direction for future research.

In the future, advancements in machine learning, deep learning, and sensor technologies are expected to further enhance robot perception capabilities. Improvements in computational power, algorithms, and data availability will enable robots to perceive and understand the world with increasing accuracy, efficiency, and adaptability. These advancements will have a profound impact on various domains, including robotics, automation, healthcare, transportation, and smart homes, paving the way for more intelligent and capable robotic systems.Stadler 1994; Bajd et al. 2010; Murphy 2014

# Chapter 6

# Robot Control

## 6.1 Introduction to Robot Control

Robot control is a crucial aspect of robotics that focuses on guiding and regulating the behavior of robots to accomplish specific tasks accurately and efficiently. Control systems play a pivotal role in enabling robots to interact with their environment, manipulate objects, navigate through space, and perform various actions with precision. Effective control is essential for ensuring the safe and reliable operation of robots across diverse domains, including manufacturing, healthcare, exploration, and service industries.

## 6.2 Control Architectures for Robots

Control architectures provide the framework for organizing and coordinating the control functions of robots. Different control architectures are employed based on the complexity of the tasks and the level of autonomy required. Here are some commonly used control architectures:

1. **Centralized Control:** In a centralized control architecture, a single controller is responsible for managing all aspects of the robot's behavior. It receives sensory inputs from various sensors, processes the information, and generates control signals to actuate the robot's actuators. Centralized control offers a unified approach for integrating different components of the robot and simplifies the implementation of complex tasks. However, it may face challenges in terms of computational load and real-time responsiveness.

2. **Decentralized Control:** Decentralized control architectures distribute control functions among multiple controllers, each handling a specific aspect of the robot's behavior. These controllers can operate independently or communicate and coordinate with each other as needed. Decentralized control provides modularity, fault tolerance, and scalability, making it suitable for systems with distributed components or multi-robot systems that require coordination among multiple entities.

3. **Hierarchical Control:** Hierarchical control architectures organize control functions into multiple levels or layers, with each level responsible for a specific aspect of the robot's behavior. Higher-level controllers handle high-level planning and decision-making, while lower-level controllers deal with low-level motion control and actuation. Hierarchical control enables the integration of complex behaviors, facilitates the separation of concerns, and allows robots to operate at different levels of autonomy, ranging from fully autonomous to teleoperated modes.

4. **Behavior-Based Control:** Behavior-based control architectures emphasize the integration of multiple concurrent behaviors to achieve desired robot actions. Each

behavior operates independently and responds to specific stimuli or conditions. The behaviors are coordinated through a selection mechanism that determines the active behavior based on the current situation. Behavior-based control offers flexibility, adaptability, and robustness in dynamic and uncertain environments, enabling robots to exhibit intelligent and reactive behavior.

# 6.3 Control Techniques for Robots

Various control techniques are employed in robot systems to achieve precise and reliable control. These techniques encompass classical control methods as well as more advanced and adaptive approaches. Here are some commonly used control techniques:

1. **Proportional-Integral-Derivative (PID) Control:** PID control is a widely used feedback control technique that utilizes the error between the desired state and the measured state to calculate control signals. The control signals adjust the robot's actuators to minimize the error and achieve the desired behavior. PID control provides stability, responsiveness, and robustness and is particularly effective for tasks requiring accurate position or velocity control, such as robot manipulators and mobile robots.

2. **Adaptive Control:** Adaptive control techniques adapt the control parameters based on changing dynamics or uncertainties in the robot system. These techniques employ online parameter estimation algorithms to continuously update the control parameters and adapt to variations in the environment or the robot's characteristics. Adaptive control enhances control performance and ro-

bustness in the presence of model uncertainties or distur-
bances, enabling robots to adapt to changing conditions.

3. **Force/Torque Control:** Force/torque control enables
   robots to interact with the environment and perform
   tasks that require compliance and force sensing capabilities. It regulates the contact forces and torques applied
   by the robot during interactions with objects or humans.

4. **Motion Planning and Trajectory Control:** Motion
   planning involves generating a sequence of robot configurations or waypoints to navigate from the current state
   to a desired goal state while avoiding obstacles. Trajectory control focuses on executing the planned trajectory
   with precision. These techniques take into account factors such as kinematics, dynamics, and constraints to
   generate smooth and collision-free motions. They enable robots to navigate complex environments and perform tasks requiring precise motion control, such as autonomous vehicles and robot manipulators.

5. **Model Predictive Control (MPC):** MPC is an advanced control technique that utilizes a predictive model
   of the robot and the environment to optimize control actions over a finite time horizon. It formulates the control
   problem as an optimization task, considering constraints,
   dynamics, and objectives. By predicting future states
   and optimizing control inputs, MPC can achieve better
   tracking performance, handle constraints effectively, and
   handle uncertain environments. MPC finds applications
   in robotics domains that require predictive and dynamic
   control, such as autonomous navigation and mobile manipulation.

6. **Machine Learning-based Control:** Machine learning
   techniques, such as neural networks, reinforcement learn-

ing, and deep learning, have gained popularity in robot control. These techniques enable robots to learn control policies directly from data or experiences, reducing the reliance on explicit models. Machine learning-based control can adapt to complex and changing environments, learn from human demonstrations, and optimize control actions based on reward signals. It has found applications in tasks like grasping and manipulation, robot locomotion, and autonomous decision-making.

7. **Hybrid Control:** Hybrid control techniques combine multiple control strategies to address different aspects of robot behavior. They integrate continuous control techniques, such as PID control or adaptive control, with discrete control modes or logical rules. Hybrid control enables robots to switch between different control strategies based on the task requirements or environmental conditions. It provides versatility and robustness in handling complex tasks that involve both continuous and discrete actions, such as human-robot collaboration and robot coordination.

8. **Safety and Fault Tolerant Control:** Safety is a critical consideration in robot control to prevent accidents and ensure human and environmental well-being. Safety control techniques focus on monitoring the robot's behavior, detecting potential hazards or anomalies, and taking appropriate actions to mitigate risks. Fault-tolerant control techniques aim to maintain robot operation even in the presence of component failures or unexpected events. These techniques employ redundancy, reconfiguration, or error-recovery mechanisms to ensure the robot's reliability and operational continuity.

In conclusion, robot control plays a pivotal role in enabling robots to perform tasks accurately and efficiently. Different control architectures and techniques are employed to achieve precise and reliable control based on the task requirements and environmental conditions. From classical control methods to advanced approaches like machine learning-based control, a wide range of techniques are available to guide the behavior of robots in diverse applications. By advancing the field of robot control, we can continue to unlock the potential of robotics and push the boundaries of what robots can accomplish in various domains.**167**; Moore et al. 2010

# Chapter 7

# Robot Programming

## 7.1 Introduction to Robot Programming

Robot programming is the process of specifying the behavior and actions of robots to perform desired tasks. It involves writing instructions or code that the robot follows to carry out its operations. Programming robots requires a combination of hardware knowledge, understanding of control algorithms, and proficiency in programming languages. Effective robot programming enables robots to exhibit intelligence, adaptability, and autonomy in their actions, making them capable of tackling complex tasks in various domains.

## 7.2 Programming Paradigms for Robots

Several programming paradigms can be applied to program robots, depending on the level of autonomy and complexity of the tasks. Here are some commonly used programming paradigms in robot programming:

1. **Sequential Programming:**  Sequential programming is the most basic paradigm, where instructions are executed one after another in a predefined order. It follows a linear flow of execution, making it suitable for simple tasks with a fixed sequence of steps. Sequential programming is often used for tasks like robot assembly or simple pick-and-place operations.

2. **Event-Driven Programming:** Event-driven programming focuses on responding to specific events or stimuli. Robots are equipped with sensors to detect changes in the environment, and when an event occurs, the robot performs the associated actions. This paradigm is useful for tasks where the robot needs to react to dynamic or unpredictable events, such as obstacle avoidance or reactive navigation.

3. **Procedural Programming:**  Procedural programming involves dividing the robot's behavior into reusable procedures or functions. Each procedure encapsulates a set of instructions that perform a specific action or subtask. Procedural programming enables modularity and code reusability, making it easier to develop and maintain complex robot behaviors. It is commonly used in applications like robot manipulation or path planning.

4. **Object-Oriented Programming (OOP):** Object-oriented programming focuses on representing the robot's behavior as a collection of objects that interact with each other. Objects encapsulate data and behaviors, and they communicate through methods or messages. OOP promotes code organization, abstraction, and encapsulation, making it easier to manage complex robot systems. It is widely used in robot control software frameworks and simulation environments.

5. **Behavior-Based Programming:** Behavior-based programming emphasizes the integration of multiple concurrent behaviors to achieve the desired robot actions. Each behavior operates independently, and their activation depends on the current situation or stimuli. Behavior-based programming enables robots to exhibit reactive and adaptive behavior, making it suitable for applications that require real-time responsiveness and robustness in dynamic environments.

# 7.3    Programming Languages for Robots

A variety of programming languages can be used to program robots, depending on the target platform, the complexity of the tasks, and the programming paradigm. Here are some commonly used programming languages in robot programming:

1. **C/C++:** C and C++ are low-level programming languages that provide direct access to hardware resources and are often used for programming embedded systems and real-time control. They offer high performance and efficiency, making them suitable for computationally intensive tasks in robotics, such as robot perception or control algorithms.

2. **Python:** Python is a high-level programming language known for its simplicity and readability. It provides extensive libraries and frameworks for robotics, making it a popular choice for rapid prototyping, algorithm development, and scripting tasks. Python is widely used in areas such as robot vision, machine learning, and robot simulation.

3. **Java:** Java is a versatile programming language known

for its platform independence and object-oriented features. It is commonly used in robotics for developing software frameworks, control systems, and user interfaces. Java's robustness and scalability make it suitable for large-scale robot systems and distributed computing environments.

4. **MATLAB:** MATLAB is a powerful programming language and environment for numerical computing and algorithm development. It provides a rich set of toolboxes and libraries specifically designed for robotics, such as Robotics System Toolbox. MATLAB is widely used for tasks such as robot kinematics and dynamics modeling, control system design, and data analysis in robotics research and development.

5. **ROS (Robot Operating System):** ROS is not a programming language itself, but a flexible framework for robot software development. It provides a set of libraries, tools, and conventions that facilitate interprocess communication, hardware abstraction, and code reusability. ROS supports multiple programming languages, including C++, Python, and Java, allowing developers to write robot software components in their preferred language. ROS has gained significant popularity in the robotics community and is widely used for developing robot control systems and coordinating multi-robot systems.

6. **Blockly and Scratch:** Blockly and Scratch are visual programming languages designed to introduce programming concepts to beginners, including children. These languages use a graphical interface where users can drag and drop blocks to create programs. Blockly and Scratch are often used in educational settings to teach robotics and programming fundamentals, enabling students to

program robots through a user-friendly and interactive interface.

7. **Simulation-specific Languages:** Simulation-specific languages, such as Gazebo's World and URDF descriptions or CoppeliaSim's Lua scripting, are used in robot simulation environments. These languages allow developers to define virtual environments, robot models, and control behaviors for testing and validation before deploying on physical robots. Simulation-specific languages provide a platform for algorithm development, testing different scenarios, and evaluating the performance of robot systems in a controlled and safe environment.

The choice of programming language depends on the specific requirements of the robot system, the development environment, and the expertise of the programming team. It is common to use a combination of programming languages and frameworks to leverage the strengths of each language and address different aspects of robot programming.

In conclusion, robot programming is a multidisciplinary field that combines hardware knowledge, control algorithms, and programming skills to specify the behavior of robots. Various programming paradigms and languages are employed to develop robot software, ranging from low-level languages like C/C++ for performance-critical tasks to high-level languages like Python for rapid prototyping and algorithm development. The emergence of frameworks like ROS has simplified the development process and facilitated code reusability and collaboration in the robotics community. As robotics continues to advance, innovative programming techniques and languages will continue to evolve, enabling robots to perform complex tasks and interact intelligently with the world around them.Siegwart, Nourbakhsh, and Scaramuzza 2011; Bhaumik 2018; Dahiya and Valle 2013

# Chapter 8

# Robot Kinematics

## 8.1 Introduction to Robot Kinematics

Robot kinematics is a branch of robotics that deals with the study of motion and geometry of robot systems. It focuses on understanding the relationship between the robot's joint angles and its resulting position and orientation in space. Kinematics plays a vital role in various aspects of robot control, including motion planning, trajectory generation, and inverse kinematics. By analyzing and modeling the kinematics of a robot, we can accurately predict its motion and enable it to perform precise tasks in its environment.

## 8.2 Forward Kinematics

Forward kinematics is the process of determining the position and orientation of a robot's end-effector based on the given joint angles. It involves computing the transformation matrix that describes the position and orientation of each link relative to the previous link in the robot's kinematic chain. The transformation matrix is obtained through the application of Denavit-Hartenberg (DH) parameters, which represent the geometric and kinematic properties of the robot's joints and links.

The DH parameters consist of four variables: link length, link twist, link offset, and joint angle. By assigning appropriate values to these parameters, we can establish a coordinate frame for each link and derive the transformation matrices using homogeneous transformations. By successively multiplying the transformation matrices from the base to the end-effector, we obtain the forward kinematic equations that relate the joint angles to the position and orientation of the end-effector.

Forward kinematics is widely used in robot manipulation tasks, where knowing the position and orientation of the end-effector is crucial for planning and executing precise movements. It is also employed in mobile robotics for determining the robot's pose in a given environment. By solving the forward kinematics equations, we can accurately position and orient the robot's end-effector, enabling it to interact with objects and navigate through space.

## 8.3 Inverse Kinematics

Inverse kinematics is the process of determining the joint angles required to achieve a desired position and orientation of the robot's end-effector. Unlike forward kinematics, which calculates the end-effector pose given the joint angles, inverse kinematics solves the problem in reverse by finding the joint angles that correspond to a specified end-effector position and orientation.

Inverse kinematics is a challenging problem due to its non-linearity and multiple solutions in certain cases. It involves solving a set of nonlinear equations that relate the joint angles to the end-effector pose. Various numerical methods, such as the Jacobian-based approaches, numerical optimization, or closed-form solutions, can be employed to solve the inverse kinematics problem.

Inverse kinematics is essential for tasks like robot manipulation, where we need to specify the desired position and orientation of the end-effector to achieve precise object grasping or tool manipulation. It is also used in motion planning to generate joint trajectories that satisfy desired constraints and avoid collisions. Inverse kinematics enables robots to perform complex tasks by providing a way to control the robot's motion based on the desired end-effector pose.

## 8.4 Differential Kinematics

Differential kinematics, also known as velocity kinematics, focuses on describing the relationship between the joint velocities and the resulting end-effector velocity. It deals with the derivatives of the forward kinematic equations and provides insights into the robot's instantaneous motion characteristics.

The differential kinematics of a robot can be derived using the Jacobian matrix, which relates the joint velocities to the end-effector velocity in Cartesian space. The Jacobian matrix provides a mapping between the joint space and the Cartesian space, allowing us to compute the end-effector velocity given the joint velocities or vice versa.

Differential kinematics is crucial for robot control tasks, such as trajectory planning, where we need to compute the joint velocities that achieve a desired end-effector position and orientation. It is also used in applications like robot teleoperation, where the robot mimics the motion of a human operator in real-time. By analyzing the differential kinematics, we can control the robot's velocity and ensure smooth and accurate movements.

## 8.5 Workspace Analysis

Workspace analysis is the study of the region in which a robot can operate and reach with its end-effector. It involves determining the boundaries, constraints, and geometric characteristics of the workspace. Understanding the robot's workspace is essential for task planning, as it helps determine whether the robot can reach the desired positions and orientations required for the task.

Workspace analysis considers factors such as joint limits, joint coupling, and physical constraints of the robot's mechanical structure. By analyzing the joint ranges and their relationships, we can define the reachable space of the robot. This analysis aids in optimizing the robot's design parameters, such as link lengths or joint limits, to maximize the workspace and improve the robot's performance.

In addition to the kinematic workspace, it is also important to consider the dynamic workspace, which takes into account factors such as joint velocities, accelerations, and dynamic limitations. Dynamic workspace analysis enables us to evaluate the robot's ability to perform tasks at different speeds and accelerations, ensuring that it can operate within desired dynamic constraints.

## 8.6 Kinematic Redundancy

Kinematic redundancy occurs when a robot has more degrees of freedom than necessary to accomplish a given task. Redundancy provides flexibility in motion and allows the robot to achieve the task objectives in multiple ways. However, it also introduces challenges in selecting the optimal joint configurations and determining the desired motion.

Kinematic redundancy can be utilized to improve the robot's performance in terms of path optimization, obstacle avoidance, or joint torque minimization. It enables the robot to adapt its joint configurations based on task requirements or environmental constraints. Various techniques, such as optimization algorithms or null-space projection methods, can be employed to exploit the redundancy and generate optimal joint trajectories.

Understanding and managing kinematic redundancy is crucial for advanced robot control and motion planning. By effectively utilizing the additional degrees of freedom, we can enhance the robot's capabilities and achieve more efficient and versatile task execution.

## 8.7 Applications of Robot Kinematics

The knowledge of robot kinematics finds application in various fields and industries. Some common applications include:

1. **Industrial Robotics:** Robot kinematics is extensively used in industrial automation for tasks such as pick-and-place operations, assembly line processes, and welding. Accurate kinematic modeling enables precise control of robot arms, ensuring efficient and reliable manufacturing processes.

2. **Medical Robotics:** In the field of medical robotics, kinematics plays a vital role in surgical robots and assistive devices. It enables precise control and manipulation

of robotic tools, facilitating minimally invasive surgeries, rehabilitation, and prosthetics.

3. **Mobile Robotics:** Robot kinematics is essential in mobile robots for navigation, mapping, and localization. By understanding the kinematics of the robot's wheels or legs, we can plan and control the robot's motion, enabling it to navigate through complex environments.

4. **Service Robotics:** Service robots, such as domestic or assistive robots, benefit from kinematic analysis to perform tasks like object manipulation, human-robot interaction, and navigation in human-centric environments. Kinematic models aid in developing intuitive and safe interaction strategies.

5. **Research and Development:** Robot kinematics is a fundamental aspect of robotics research and development. It provides the foundation for designing new robot architectures, developing advanced control algorithms, and exploring novel robot applications.

In conclusion, robot kinematics is a fundamental discipline in robotics that deals with the study of motion and geometry of robot systems. Understanding the forward and inverse kinematics, differential kinematics, workspace analysis, and kinematic redundancy is essential for accurate robot control, motion planning, and task execution. By gaining insights into the kinematics of robots, we can design more efficient and capable robotic systems that can perform a wide range of tasks in various industries and domains.

Furthermore, advancements in robot kinematics have led to significant progress in fields such as human-robot collaboration, where robots work alongside humans in shared workspaces. By understanding the kinematic constraints and capabilities of robots, we can ensure safe and effective collaboration between humans and robots, enhancing productivity and expanding the possibilities of automation.

In recent years, there have been notable developments in the field of soft robotics, which involves designing and controlling robots with compliant and flexible structures. Soft robots exhibit unique kinematic properties that differ from traditional rigid-body robots. The study of kinematics in soft robotics enables researchers to understand the deformation and motion of soft robot bodies, opening up new avenues for applications in areas such as medical robotics, exploration, and human interaction.

The future of robot kinematics holds promising opportunities for advancements in areas such as robot learning, where robots can autonomously acquire and refine their kinematic models through interaction with their environment. This can lead to more adaptive and intelligent robots capable of adapting to novel situations and performing complex tasks with minimal human intervention.

As the field of robotics continues to evolve, a deep understanding of robot kinematics will remain critical for researchers, engineers, and practitioners. It will enable them to develop sophisticated robot systems that can navigate and manipulate their environment with precision, interact with humans in a natural and intuitive manner, and contribute to various industries, from manufacturing to healthcare and beyond.

In conclusion, this chapter has provided an overview of robot kinematics, including forward and inverse kinematics, differential kinematics, workspace analysis, kinematic redundancy, and their applications. By studying the motion and geometry of robots, we can unlock their full potential and create advanced robotic systems that enhance productivity, improve safety, and transform various industries. Robot kinematics serves as a foundation for further research and development in the field of robotics, paving the way for exciting advancements in the future.Siegwart, Nourbakhsh, and Scaramuzza 2011; Bhaumik 2018

# Chapter 9

# Robot Sensors

## 9.1 Introduction to Robot Sensors

Robot sensors are essential components that enable robots to perceive and interact with their environment. They provide robots with the ability to sense and gather information about their surroundings, allowing them to make informed decisions and carry out tasks effectively. Sensors act as the robot's senses, providing inputs for various processes such as navigation, object detection, localization, and feedback control.

There are numerous types of sensors used in robotics, each serving a specific purpose and providing unique capabilities. These sensors can be categorized into different modalities, including:

1. **Vision Sensors:** Vision sensors, such as cameras and depth sensors, enable robots to perceive the visual information from their surroundings. They capture images or depth maps, which can be processed to recognize objects, estimate distances, and navigate the environment. Vision sensors are widely used in applications like object detection, tracking, mapping, and visual servoing.

2. **Tactile Sensors:** Tactile sensors provide robots with a sense of touch, allowing them to detect and measure physical contact with objects or surfaces. These sensors can range from simple contact switches to more advanced technologies such as force-sensitive resistors or tactile arrays. Tactile sensing is crucial for tasks such as object manipulation, grasping, and interaction with humans.

3. **Range Sensors:** Range sensors, including ultrasonic sensors, laser rangefinders, and LiDAR (Light Detection and Ranging) sensors, provide robots with distance measurements to objects in their vicinity. These sensors emit a signal and measure the time it takes for the signal to bounce back, allowing the robot to estimate distances and generate a map of the environment. Range sensors are commonly used in mapping, obstacle avoidance, and localization.

4. **Proximity Sensors:** Proximity sensors detect the presence or absence of objects in close proximity to the robot. They can utilize various technologies such as infrared, capacitive, or inductive sensing. Proximity sensors are often employed in collision avoidance systems, robot safety, and human-robot interaction.

5. **Inertial Sensors:** Inertial sensors, including accelerometers and gyroscopes, measure the acceleration and angular velocity of the robot. These sensors provide information about the robot's orientation, motion, and changes in velocity. Inertial sensors are essential for tasks such as motion control, balance control, and inertial navigation.

6. **Environmental Sensors:** Environmental sensors, such as temperature sensors, humidity sensors, and gas sensors, enable robots to gather information about the environmental conditions. These sensors are useful in appli-

cations like environmental monitoring, indoor air quality assessment, and autonomous systems that need to adapt to their surroundings.

7. **Force/Torque Sensors:** Force/torque sensors measure the forces and torques applied to the robot's end-effector or joints. These sensors provide valuable feedback for tasks requiring force control, such as manipulation, assembly, or compliance control. Force/torque sensors can detect contact forces, assess object stiffness, and enable delicate interactions with the environment.

The choice of sensors depends on the specific requirements of the robot's tasks and the characteristics of the environment in which it operates. In many cases, robots use a combination of different sensors to gather multiple types of information and create a comprehensive understanding of their surroundings.

## 9.2   Sensor Integration and Fusion

Sensor integration and fusion involve combining information from multiple sensors to enhance the robot's perception capabilities and improve the reliability and accuracy of the data. By fusing data from different sensors, robots can overcome the limitations of individual sensors and obtain a more comprehensive understanding of the environment.

Sensor fusion can be performed at different levels, including:

1. **Data-Level Fusion:** Data-level fusion involves combining raw sensor data to generate a unified representation. This can be achieved through techniques such as sensor calibration, time synchronization, data alignment, and data correlation. By merging the raw data, the robot

can obtain a more accurate and complete representation of the environment.

2. **Feature-Level Fusion:** Feature-level fusion involves extracting relevant features from each sensor's data and combining them to create a higher-level representation. This can include features such as object recognition, motion estimation, or spatial mapping. By integrating the extracted features, the robot can have a more robust perception of its surroundings.

3. **Decision-Level Fusion:** Decision-level fusion focuses on combining the outputs or decisions made by individual sensors to make a final decision or inference. This can involve techniques such as voting systems, weighted averaging, or Bayesian inference. Decision-level fusion enables the robot to make informed decisions based on the collective information provided by the sensors.

Sensor integration and fusion techniques play a crucial role in improving the reliability and accuracy of robot perception. By combining the strengths of different sensors and mitigating their individual limitations, robots can achieve more robust and adaptable perception capabilities.

# 9.3   Sensor Calibration and Calibration Techniques

Sensor calibration is the process of determining and minimizing the errors or inaccuracies associated with sensor measurements. Calibration is essential to ensure that the sensor readings correspond accurately to the physical quantities they are intended to measure. It involves estimating the systematic errors, such as bias, scale factor, or non-linearity, and compensating for them to improve measurement accuracy.

There are various calibration techniques depending on the type of sensor:

1. **Camera Calibration:** Camera calibration involves estimating the intrinsic and extrinsic parameters of a camera to correct distortions and accurately map 2D image coordinates to 3D world coordinates. Techniques like the Zhang's method or the Tsai's method are commonly used for camera calibration.

2. **Range Sensor Calibration:** Range sensors, such as LiDAR or depth sensors, require calibration to ensure accurate distance measurements. Calibration involves determining the sensor's pose and intrinsic parameters, as well as compensating for systematic errors like sensor noise or misalignment.

3. **Inertial Sensor Calibration:** Inertial sensors, such as accelerometers and gyroscopes, need calibration to remove biases and ensure accurate measurement of linear accelerations and angular velocities. Calibration techniques involve static or dynamic calibration methods, where the sensor is subjected to known motion patterns or placed in a stable environment to estimate the calibration parameters.

4. **Force/Torque Sensor Calibration:** Force/torque sensors require calibration to accurately measure forces and torques. Calibration involves applying known forces or torques to the sensor and comparing them with the sensor readings to estimate calibration parameters such as stiffness, scale factor, or misalignment.

Calibration is an iterative process that involves collecting data, estimating calibration parameters, and refining the calibration model. Accurate sensor calibration is crucial for reliable robot perception and control, as it ensures that the measurements align with the real-world physical quantities.

## 9.4   Sensor Fusion for Localization and Mapping

Sensor fusion plays a vital role in simultaneous localization and mapping (SLAM) algorithms. SLAM is a fundamental problem in robotics, where a robot explores an unknown environment, builds a map of the environment, and simultaneously estimates its own pose or location within that map.

By integrating data from different sensors, such as range sensors, vision sensors, and odometry sensors, SLAM algorithms can generate more accurate maps and estimate the robot's pose with higher precision. Range sensors provide distance measurements to objects, vision sensors provide visual features for localization and mapping, and odometry sensors estimate the robot's motion based on wheel encoders or inertial sensors.

Sensor fusion techniques, such as extended Kalman filters (EKF) or particle filters, are commonly used in SLAM to fuse the sensor data, estimate the robot's pose and map the environment. These algorithms incorporate measurements from multiple sensors, handle uncertainties, and provide a consistent and accurate representation of the robot's position and the environment.Dahiya and Valle 2013; Faust 2007

# Chapter 10

# Robot Vision

## 10.1 Introduction to Robot Vision

Robot vision is a branch of robotics that focuses on the development and application of visual sensing and processing capabilities in robots. Vision plays a crucial role in enabling robots to perceive, understand, and interact with their environment. By incorporating vision systems, robots can extract meaningful information from visual inputs, such as images or video streams, and use that information to make informed decisions and perform complex tasks.

The field of robot vision encompasses various sub-disciplines, including image processing, pattern recognition, computer vision, and machine learning. These disciplines provide the foundation for developing algorithms and techniques that enable robots to interpret visual data, recognize objects, estimate poses, track motion, and navigate in their surroundings.

## 10.2 Image Formation and Processing

Image formation is the process by which a camera captures the visual information from the scene and converts it into a digital representation. Understanding image formation is essential for robot vision, as it helps in interpreting and processing the acquired images effectively.

The main components involved in image formation are:

Optics: The optics of a camera system determine how light rays from the scene are focused onto the image sensor. The lens plays a critical role in capturing the desired field of view, controlling depth of field, and minimizing optical distortions.

1. **Image Sensor:** The image sensor, typically a CCD or CMOS sensor, converts the incoming light into electrical signals. Each pixel on the sensor captures the intensity and color information of the corresponding portion of the scene.

2. **Image Processing:** Image processing techniques involve manipulating and enhancing the acquired images to extract relevant features or information. This includes operations such as noise reduction, image enhancement, edge detection, image segmentation, and feature extraction.

Image processing algorithms are essential for preprocessing the images before higher-level tasks, such as object recognition or tracking, can be performed.

# 10.3 Object Detection and Recognition

Object detection and recognition are fundamental tasks in robot vision, enabling robots to identify and locate objects of interest within the visual scene. These tasks are crucial for various robotic applications, such as pick-and-place operations, object manipulation, and scene understanding.

Object detection involves localizing and identifying instances of objects within an image or a video stream. It typically involves using techniques such as feature extraction, classification, and localization. Object recognition, on the other hand, focuses on identifying the object category or type based on its visual appearance.

There are several approaches to object detection and recognition, including:

1. **Traditional Feature-based Methods:** These methods rely on extracting handcrafted features from images, such as edges, corners, or texture descriptors. These features are then used to train classifiers, such as support vector machines (SVM) or random forests, to distinguish between different object classes.

2. **Deep Learning-based Methods:** Deep learning has revolutionized object detection and recognition in recent years. Convolutional Neural Networks (CNNs) have shown remarkable performance in detecting and recognizing objects in images. Networks such as Faster R-CNN, YOLO (You Only Look Once), and SSD (Single Shot MultiBox Detector) have achieved state-of-the-art results in real-time object detection

3. **3D Object Recognition:** In addition to 2D object recognition, there is a growing interest in 3D object

recognition, where robots can recognize objects based on their 3D geometry and appearance. This involves techniques such as 3D point cloud processing, depth image analysis, or the fusion of 2D and 3D information.

Object detection and recognition algorithms are essential for enabling robots to interact with their environment autonomously and perform tasks that require understanding and manipulation of objects.

# 10.4 Robot Navigation and Visual SLAM

Robot navigation is the ability of a robot to move autonomously in its environment, avoiding obstacles, and reaching desired destinations. Visual perception plays a crucial role in robot navigation, as it enables the robot to understand and interpret the surrounding environment to plan its path.

Visual Simultaneous Localization and Mapping (SLAM) is a key technique used in robot navigation. SLAM allows a robot to simultaneously build a map of its environment and estimate its own position within that map using visual inputs. Visual SLAM algorithms utilize the visual information captured by cameras to reconstruct the environment and localize the robot within it.

There are several approaches to visual SLAM:

1. **Feature-based SLAM:** This approach involves extracting visual features, such as keypoints or landmarks, from the images and using them to build a map. The robot estimates its position by matching the observed features with the ones stored in the map. Feature-based SLAM algorithms include methods like ORB-SLAM, PTAM (Par-

allel Tracking and Mapping), and SVO (Semi-Direct Visual Odometry).

2. **Direct SLAM:** Direct SLAM methods utilize the raw intensity or color information from the images to directly estimate the camera's pose and reconstruct the environment. These methods avoid feature extraction and matching steps and operate directly on the pixel intensities. Examples of direct SLAM algorithms include DTAM (Dense Tracking and Mapping) and LSD-SLAM (Large-Scale Direct SLAM).

3. **Semantic SLAM:** Semantic SLAM aims to not only reconstruct the environment but also understand the semantic meaning of the observed scenes. It involves recognizing objects, segmenting the scene into meaningful regions, and incorporating semantic information into the mapping and localization process. This enables the robot to have a higher-level understanding of its environment. Semantic SLAM combines visual information with techniques from object recognition and semantic segmentation.

4. **Visual SLAM :** Visual SLAM algorithms provide robots with the capability to navigate in unknown environments, localize themselves accurately, and create detailed maps that can be used for path planning and obstacle avoidance. By leveraging visual perception, robots can navigate more efficiently and adapt to dynamic environments.

# 10.5 Robot Interaction and Human-Robot Collaboration

Robot vision also plays a crucial role in enabling robots to interact and collaborate with humans. By perceiving and understanding human gestures, expressions, and actions, robots can effectively communicate and collaborate with humans in various settings.

Human-Robot Interaction (HRI) encompasses a range of capabilities, including:

1. **Gesture Recognition:** Robots can interpret human gestures and body movements to understand intentions and commands. This enables natural and intuitive communication between humans and robots.

2. **Facial Expression Analysis:** By analyzing facial expressions, robots can infer human emotions and respond accordingly. This capability is particularly important in social robots or robots designed to provide assistance and support.

3. **Object Recognition and Grasping:** Robots with vision capabilities can recognize and manipulate objects in collaboration with humans. They can understand human instructions regarding object manipulation tasks and execute them accordingly.

4. **Scene Understanding:** Robots equipped with vision systems can understand and interpret the scene context, including objects, activities, and spatial relationships. This allows them to adapt their behavior and actions to the given situation.

By integrating vision-based interaction capabilities, robots can effectively collaborate with humans in various domains, including healthcare, manufacturing, entertainment, and personal assistance.

In conclusion, robot vision plays a critical role in enabling robots to perceive, understand, and interact with their environment. Through image formation, processing, object detection and recognition, navigation, and human-robot collaboration, robots equipped with vision systems can navigate autonomously, manipulate objects, and collaborate with humans effectively. The advancements in computer vision, machine learning, and robotics have paved the way for more capable and intelligent robots that can operate in complex and dynamic real-world scenarios.Vukobratovic and Stokic 2012; Mittal et al. 1998

# Chapter 11

# Robot Communication

## 11.1 Introduction to Robot Communication

Robot communication refers to the exchange of information between robots, as well as between robots and humans or other external entities. Communication is a crucial aspect of robotics, as it enables robots to share data, coordinate actions, and interact with their surroundings effectively. By establishing reliable and efficient communication channels, robots can enhance their capabilities and operate in collaborative environments.

Robot communication can be categorized into two main types:

1. **Inter-Robot Communication:** This type of communication involves the exchange of information between multiple robots. Inter-robot communication enables coordination and collaboration among robots, allowing them to work together towards a common goal. Examples of inter-robot communication include sharing sensor data,

coordinating movements, task allocation, and sharing environmental maps.

2. **Human-Robot Communication:** Human-robot communication focuses on the interaction and communication between robots and humans. It enables humans to convey instructions, commands, or queries to robots, and allows robots to provide feedback, updates, or requests to humans. Human-robot communication can take various forms, including speech, gestures, touch, visual cues, or a combination of these modalities.

## 11.2 Communication Technologies for Robots

There are several communication technologies and protocols used in robot communication, depending on the requirements of the application and the environment. Some common communication technologies for robots include:

1. **Wireless Communication:** Wireless communication enables robots to exchange data without the need for physical connections. Wi-Fi, Bluetooth, and Zigbee are commonly used wireless communication protocols in robotics. They provide reliable and high-speed data transmission, allowing robots to communicate with each other or with external devices such as smartphones, computers, or control systems.

2. **Wired Communication:** In some cases, wired communication is preferred for its reliability and security. Ethernet is a widely used wired communication protocol in robotics, offering high-speed and robust data transmission. Wired communication is often used in indus-

trial settings or scenarios where the environment is prone to electromagnetic interference or wireless signal limitations.

3. **CAN Bus:** Controller Area Network (CAN) is a serial communication bus commonly used in robotics and automation systems. CAN bus allows for reliable and real-time communication between different devices, sensors, and actuators. It is particularly suitable for distributed control systems where multiple robots or subsystems need to communicate and synchronize their actions.

4. **ROS (Robot Operating System):** ROS is a popular framework used in robotics for communication and integration of various components. ROS provides a flexible and standardized communication infrastructure that enables different modules and nodes to exchange messages, commands, and data. It simplifies the development of robot systems by providing a common communication protocol and a wide range of tools and libraries.

5. **Internet-based Communication:** With the advent of the Internet of Things (IoT), robots can leverage internet-based communication protocols to connect with cloud services, access databases, or communicate with remote users. MQTT (Message Queuing Telemetry Transport) and RESTful APIs (Representational State Transfer) are examples of protocols used for internet-based communication in robotics.

The choice of communication technology depends on factors such as bandwidth requirements, latency constraints, power consumption, security concerns, and the specific application context.

# 11.3   Communication Protocols and Message Formats

To enable effective communication, robots utilize various communication protocols and message formats to exchange information in a structured and standardized manner. These protocols define rules and formats for data transmission and ensure compatibility between different robotic systems.

Some commonly used communication protocols in robotics include:

1. **TCP/IP (Transmission Control Protocol/Internet Protocol):** TCP/IP is the foundational protocol of the internet and provides a reliable and connection-oriented communication between devices. It is often used for high-level communication between robots and external systems, such as control centers or user interfaces.

2. **UDP (User Datagram Protocol):** UDP is a connectionless and lightweight communication protocol. Unlike TCP/IP, UDP does not guarantee reliable delivery or ordering of messages. However, it offers lower latency and is well-suited for real-time applications where speed is crucial. UDP is often used for low-level communication between robots and for transmitting sensor data that requires fast updates, such as in real-time control systems or time-critical operations.

3. **MQTT (Message Queuing Telemetry Transport):** MQTT is a lightweight publish-subscribe messaging protocol designed for constrained devices and low-bandwidth networks. It follows a publish-subscribe model, where robots can publish messages to specific topics, and other robots or systems can subscribe to those topics to receive the messages. MQTT is commonly used in IoT

applications, allowing robots to communicate efficiently with other devices and cloud services.

4. **HTTP (Hypertext Transfer Protocol):** HTTP is a widely used protocol for communication between web browsers and servers. It is based on a client-server architecture, where robots can act as either clients or servers. HTTP provides a standardized format for requesting and exchanging data, making it suitable for web-based interfaces and remote control of robots through web browsers or APIs.

5. **JSON (JavaScript Object Notation):** JSON is a lightweight and human-readable data interchange format. It is widely used for representing structured data in web-based communication and APIs. JSON provides a flexible format for encoding and decoding data, allowing robots to exchange information in a platform-independent manner.

6. **XML (eXtensible Markup Language):** XML is a markup language that provides a hierarchical structure for representing data. It is widely used for data exchange and configuration in robotic systems. XML allows for defining custom data structures and schemas, making it suitable for complex communication scenarios where data validation and extensibility are important.

7. **ROS (Robot Operating System):** ROS, mentioned earlier in this chapter, utilizes its own communication protocol called the ROS message format. ROS messages define the structure and content of the data exchanged between different ROS nodes. The ROS communication infrastructure enables seamless integration and communication between various modules and components within a robotic system.

It is worth noting that the choice of communication protocol and message format depends on factors such as the complexity of the data to be exchanged, the bandwidth requirements, the desired level of reliability, and the compatibility with existing systems or frameworks.

## 11.4    Communication Challenges and Considerations

Robot communication faces several challenges and considerations that need to be addressed for reliable and effective operation:

1. **Bandwidth and Latency:** Robots often generate and exchange large amounts of data, including sensor readings, images, and control commands. Ensuring sufficient bandwidth and low latency in communication channels is crucial, especially for real-time applications and time-sensitive operations.

2. **Network Connectivity:** Robots may operate in environments with limited or intermittent network connectivity, such as remote locations or areas with weak wireless signals. Communication protocols should be designed to handle such scenarios, allowing robots to adapt to varying network conditions and recover from connection failures.

3. **Security:** Robot communication may involve sensitive data or control commands, making security a critical consideration. Encryption, authentication, and access control mechanisms should be implemented to protect data integrity, confidentiality, and prevent unauthorized access or tampering.

4. **Scalability:** As robotic systems grow in complexity and the number of robots involved increases, communication protocols should be scalable to handle the growing network size and message traffic. Efficient message routing, load balancing, and network management techniques are important for maintaining robust communication in large-scale deployments.

5. **Interoperability:** Robots often need to communicate with external systems, such as control centers, databases, or other robots from different manufacturers. Interoperability standards and protocols should be employed to ensure seamless communication and integration between heterogeneous systems.

6. **Error Handling and Reliability:** Communication errors, packet loss, and network disruptions are common in dynamic environments. Robust error handling mechanisms should be in place to detect and recover from errors, ensuring reliable and fault-tolerant communication. Retransmission protocols, error correction codes, and message acknowledgment techniques can be employed to improve reliability.

7. **Quality of Service (QoS):** Some robotic applications require specific quality of service guarantees, such as low latency, high reliability, or prioritized message delivery. QoS mechanisms can be implemented in communication protocols to provide differentiated treatment for different types of data, ensuring that critical messages receive timely and guaranteed delivery.

8. **Synchronization and Time Stamping:** In multi-robot systems or systems with distributed components, synchronization of clocks and time stamping of messages

are important for accurate coordination and data fusion. Time synchronization protocols, such as the Network Time Protocol (NTP) or Precision Time Protocol (PTP), can be used to achieve consistent time references across robots and systems.

9. **Human-Robot Interaction:** In human-robot communication, the design of user interfaces, feedback mechanisms, and natural language processing plays a vital role. Robots should be able to understand and interpret human commands, gestures, or speech, and provide appropriate feedback or responses to ensure effective collaboration and interaction.

10. **Ethical and Social Considerations:** As robots become more integrated into society, ethical and social considerations in communication become crucial. Privacy, data protection, consent, and transparency should be addressed to build trust and ensure responsible use of robot communication technologies.

# 11.5 Future Trends in Robot Communication

Robot communication continues to evolve, driven by advancements in technology and the increasing complexity of robotic systems. Some future trends and directions in robot communication include:

1. **5G and Beyond:** The deployment of 5G networks and upcoming 6G networks will revolutionize robot communication. With ultra-high-speed data transmission, low latency, and massive device connectivity, robots will be

able to communicate and collaborate in real-time, enabling new applications and services.

2. **Edge Computing:** Edge computing brings computation and data processing closer to the robots, reducing latency and dependency on cloud services. Edge-based communication and computation allow robots to make real-time decisions and exchange information locally, improving responsiveness and autonomy.

3. **Multi-Modal Communication:** Robots will increasingly leverage multiple communication modalities, such as speech, gestures, vision, and haptic interfaces, to enhance human-robot interaction and enable more natural and intuitive communication.

4. **Artificial Intelligence (AI) in Communication:** AI techniques, such as natural language processing, machine learning, and computer vision, will play a significant role in improving robot communication. AI can enhance speech recognition, enable context-aware understanding of commands, and enable robots to generate human-like responses.

5. **Swarm Robotics:** Swarm robotics involves the coordination and communication of large groups of robots to accomplish tasks collectively. Communication protocols for swarm robotics focus on self-organization, scalability, and decentralized decision-making, allowing robots to work together as a cohesive group.

6. **Blockchain Technology:** Blockchain, with its decentralized and secure nature, has the potential to enhance trust, data integrity, and privacy in robot communication. It can provide an immutable record of communication events and facilitate secure peer-to-peer transactions between robots.

7. **Socially-Aware Communication:** As robots become more integrated into human social environments, communication protocols will consider social norms, cultural differences, and ethical guidelines to enable robots to interact respectfully and effectively with humans.

In conclusion, robot communication is a fundamental aspect of robotics, enabling robots to exchange information, collaborate, and interact with humans and other robots. Various communication technologies, protocols, and message formats are employed to ensure reliableDavies 1997; Liljebäck et al. 2013

# Chapter 12

# Robot Planning

## 12.1 Introduction to Robot Planning

Robot planning is a fundamental aspect of autonomous robotic systems, enabling them to generate sequences of actions that accomplish desired goals in complex and dynamic environments. Planning involves determining the appropriate actions to achieve specific objectives, considering the robot's capabilities, the environment, and any constraints or uncertainties.

In this chapter, we will explore the principles, techniques, and algorithms involved in robot planning. We will discuss different planning paradigms, such as classical planning, motion planning, and task planning. Additionally, we will delve into the challenges, considerations, and real-world applications of robot planning.

## 12.2 Classical Planning

Classical planning focuses on generating a sequence of actions in a deterministic, discrete world. It operates under the assumption that the robot has complete knowledge of the environment and can accurately predict the effects of its actions. Classical planning problems are typically formulated using formal languages such as the Planning Domain Definition Language (PDDL) or STRIPS.

The process of classical planning involves defining the initial state, goal state, and a set of actions with preconditions and effects. The planner then employs search algorithms, such as A*, breadth-first search, or depth-first search, to explore the space of possible action sequences and find a solution that reaches the desired goal state.

Planning algorithms in classical planning can handle various complexities, including constraints, time, resources, and uncertainty. Techniques like partial-order planning, hierarchical planning, and plan repair enable more flexible and efficient planning in dynamic environments.

## 12.3   Motion Planning

Motion planning deals with the problem of finding a collision-free path for a robot in continuous space. Unlike classical planning, motion planning considers the physical constraints and dynamics of the robot, as well as obstacles in the environment. It aims to determine the robot's trajectory that ensures safe and efficient movement.

Motion planning algorithms utilize techniques such as sampling-based methods, graph search, potential fields, or optimization-based approaches to explore the configuration space and generate feasible paths. Algorithms like Rapidly-exploring Random Trees (RRT) and A* with heuristics are commonly employed in motion planning.

Real-time motion planning is essential for robots operating in dynamic environments where obstacles or other agents' positions may change rapidly. Adaptive and reactive planning techniques enable robots to dynamically adjust their paths to avoid collisions and respond to changing circumstances.

## 12.4   Task Planning

Task planning focuses on high-level decision-making and sequencing of actions to achieve complex goals involving multiple subtasks. Task planning takes into account the temporal and logical dependencies among actions and considers factors such as resource allocation, deadlines, and constraints.

Task planning often involves task decomposition, where complex tasks are divided into subtasks and organized into a hierarchical structure. Planning algorithms, such as Hierarchical Task Network (HTN) planning or Temporal Planning Networks (TPN), are used to generate a plan that satisfies the overall objectives and constraints.

Task planning can incorporate uncertainty and adaptability through techniques like plan repair, plan merging, or plan revision. These approaches enable robots to handle unforeseen events, changes in the environment, or failures during execution.

## 12.5   Challenges and Considerations in Robot Planning

Robot planning faces several challenges that need to be addressed for successful implementation:

1. **State Estimation and Perception:** Accurate perception of the environment and estimation of the robot's

state are crucial for effective planning. Perception techniques, such as computer vision, sensor fusion, or localization algorithms, provide the necessary inputs for planning algorithms to generate reliable plans.

2. **Scalability and Efficiency:** Planning algorithms should scale well with the complexity of the problem and the size of the environment. Efficient data structures, heuristic guidance, parallelization, or approximation techniques help improve planning speed and scalability.

3. **Uncertainty and Sensing Limitations:** Robots often operate in uncertain and dynamic environments, where sensing limitations and uncertainty in perception can pose challenges to planning. Techniques like probabilistic planning, Bayesian inference, or Monte Carlo methods can be employed to handle uncertainty and make informed decisions.

4. **Real-time Planning:** In time-critical scenarios, robots need to generate plans and make decisions quickly. Real-time planning algorithms, efficient heuristics, and parallel processing can enable robots to plan and react in real-time, ensuring timely and responsive behavior.

5. **Integration with Execution and Control:** Planning and execution should be tightly integrated to ensure that planned actions are effectively executed. Feedback loops, monitoring, and error handling mechanisms bridge the gap between planning and execution, allowing robots to adapt and handle unexpected situations during plan execution.

6. **Human-Robot Collaboration:** In scenarios where humans and robots work together, planning should consider human preferences, intentions, and safety constraints.

Collaborative planning techniques enable robots to generate plans that accommodate human inputs, respect safety requirements, and ensure smooth collaboration.

# 12.6  Real-World Applications of Robot Planning

Robot planning has diverse applications across various domains, including:

1. **Autonomous Vehicles:** Planning plays a crucial role in autonomous vehicle navigation, route planning, and collision avoidance. Planning algorithms help vehicles generate safe and efficient trajectories while considering traffic rules, obstacles, and dynamic changes in the environment.

2. **Warehouse Automation:** Planning is essential for efficient warehouse operations, where robots need to navigate through cluttered environments, plan optimal paths for item retrieval, and coordinate with other robots to optimize task allocation and order fulfillment.

3. **Manufacturing and Assembly:** Planning enables robots to perform complex assembly tasks by determining optimal grasping strategies, motion paths, and task sequencing. Planning algorithms ensure efficient resource utilization, reduce errors, and improve productivity in manufacturing processes.

4. **Search and Rescue:** In search and rescue scenarios, robots can autonomously plan their paths to explore unknown or hazardous environments, locate survivors, and

navigate obstacles. Planning algorithms aid in optimizing search strategies and coordinating multiple robots for efficient rescue operations.

5. **Environmental Monitoring:** Robots equipped with sensing capabilities can plan their paths to collect environmental data, monitor pollution levels, or perform surveillance. Planning algorithms enable efficient coverage of the target area, adaptive sampling, and data gathering in environmental monitoring applications.

6. **Service Robotics:** Planning is crucial in service robots that assist with tasks such as cleaning, delivery, or assistance to humans. Robots need to plan their motions, navigate around obstacles, and interact with objects or humans in a safe and efficient manner.

In conclusion, robot planning is a vital component of autonomous robotic systems, enabling them to generate sequences of actions that achieve desired objectives in complex and dynamic environments. Whether it is classical planning, motion planning, or task planning, the selection of appropriate planning techniques depends on the specific requirements and constraints of the application. Overcoming challenges such as uncertainty, real-time constraints, and integration with execution ensures the successful implementation of planning algorithms in various real-world applications across different domains.Murray et al. 1994; Fairhurst 1988

# Chapter 13

# Robot Applications

## 13.1 Introduction to Robot Applications

Robots have revolutionized various industries and sectors, offering solutions to complex tasks, improving efficiency, and enhancing human safety. In this chapter, we will explore the wide range of applications where robots are making significant contributions. From manufacturing and healthcare to exploration and entertainment, robots are playing an increasingly vital role in diverse domains.

## 13.2 Manufacturing and Industrial Robotics

One of the most prominent areas where robots have made a significant impact is in manufacturing and industrial automation. Robots have revolutionized assembly lines, enabling faster and more precise production processes. They can perform repetitive tasks with high accuracy, increasing productivity and reducing human error. Industrial robots are used for tasks such as welding, painting, material handling, and quality inspection. Collaborative robots, also known as cobots, work alongside human workers, enhancing safety and efficiency in manufacturing environments.

## 13.3 Healthcare and Medical Robotics

Robotics is transforming the healthcare industry, offering innovative solutions for diagnosis, treatment, and patient care. Surgical robots enable minimally invasive procedures with enhanced precision, reducing patient trauma and recovery time. Robotic exoskeletons assist in rehabilitation and mobility enhancement for individuals with physical impairments. Autonomous robotic systems are employed in healthcare facilities for tasks such as medication delivery, disinfection, and patient monitoring. Robots also play a role in telemedicine, enabling remote consultations and remote surgery, expanding access to healthcare services.

## 13.4 Agriculture and Farming Robotics

In the agricultural sector, robots are being used to automate various tasks, improving efficiency and reducing labor costs. Agricultural robots can perform activities such as planting, harvesting, and crop monitoring. They can autonomously navigate fields, identify and remove weeds, and optimize irrigation and fertilization processes. Robotic systems equipped with sensors and computer vision can analyze crop health, detect diseases, and assist in precision agriculture practices. By enabling more efficient resource utilization, robots contribute to sustainable farming practices.

# 13.5   Exploration and Space Robotics

Robots play a crucial role in space exploration, assisting in the exploration of planets, moons, and celestial bodies. Robotic rovers, like NASA's Mars rovers, are designed to navigate and investigate extraterrestrial environments, collect samples, and transmit data back to Earth. Space probes and satellites employ robotic systems for tasks such as orbit correction, instrument deployment, and maintenance. Robots enable scientific discoveries and expand our understanding of the universe while overcoming the challenges of harsh and remote environments.

# 13.6   Logistics and Warehouse Robotics

Logistics and warehouse operations have been revolutionized by the use of robots. Autonomous mobile robots navigate warehouse environments, efficiently picking and placing items, optimizing order fulfillment processes, and reducing human effort. Robots collaborate with human workers to streamline inventory management, reduce errors, and enhance the overall efficiency of logistics operations. Robotic systems also play a role in last-mile delivery, with autonomous delivery robots and drones being deployed for package transportation.

# 13.7   Service Robotics

Service robots are designed to interact and assist humans in various settings. They have applications in sectors such as hospitality, healthcare, retail, and entertainment. Social robots provide companionship and support to individuals, particularly the elderly and people with special needs. Service robots can perform tasks such as cleaning, customer assistance, information provision, and entertainment. They enhance customer experiences, improve operational efficiency, and contribute to the well-being of individuals in diverse environments.

## 13.8   Education and Research Robotics

Robots are increasingly being used as educational tools to inspire and engage students in science, technology, engineering, and mathematics (STEM) disciplines. Educational robots enable hands-on learning experiences, teaching programming, problem-solving, and robotics concepts in an interactive manner. Research robotics involves the development of advanced robotic systems and technologies to explore new possibilities and push the boundaries of robotics research. Researchers use robots to study human-robot interaction, machine learning algorithms, sensor integration, and advanced control strategies. Research in robotics contributes to the development of new robotic applications and drives innovation in various fields.

## 13.9   Entertainment and Recreation Robotics

Robots have found their way into the entertainment industry, bringing joy and entertainment to people of all ages. Entertainment robots include robotic toys, interactive companions, and humanoid robots that can dance, sing, and engage in interactive activities. Theme parks and entertainment venues employ robots for attractions, shows, and immersive experiences. Robotic technology adds a new dimension of excitement and engagement, captivating audiences and creating memorable experiences.

## 13.10   Environmental and Hazardous Tasks Robotics

Robots are utilized in environments that are hazardous or inaccessible to humans. They can perform tasks such as disaster response, hazardous material handling, and inspection of dangerous structures. Robotic systems equipped with sensors and cameras can assess hazardous situations, detect gas leaks, and remotely operate in dangerous environments. By minimizing human exposure to risk, robots contribute to improved safety and efficiency in high-risk scenarios.

## 13.11   Defense and Security Robotics

The defense and security sector utilize robots for various purposes, including reconnaissance, surveillance, bomb disposal, and mine detection. Unmanned aerial vehicles (UAVs) or drones are used for aerial surveillance and monitoring. Ground-based robots navigate challenging terrains and perform tasks such as detecting and neutralizing explosives. Robots assist in border control, disaster response, and maintaining public safety, providing valuable support to security forces.

## 13.12   Personal and Domestic Robotics

Robotic technology is also making its way into homes and personal environments. Personal robots can assist with household chores, such as vacuuming, mopping, and lawn mowing. Smart home systems integrate robotic devices for automation and control of various tasks, enhancing convenience and comfort. Robots for elder care can provide assistance with daily activities, medication reminders, and remote monitoring, enabling aging individuals to maintain independence and quality of life.

## 13.13    Ethical, Legal, and Social Considerations

As robotics technology advances and robots become more integrated into our daily lives, it is essential to address ethical, legal, and social considerations. Discussions around robot autonomy, job displacement, privacy, safety, and ethical decision-making algorithms are crucial for responsible and beneficial deployment of robots in society. It is important to ensure that robots are designed and used in ways that align with societal values, respect human rights, and prioritize human well-being.

In conclusion, robots have a vast range of applications across diverse sectors, transforming industries, and improving various aspects of human life. Whether it is in manufacturing, healthcare, exploration, entertainment, or personal settings, robots are contributing to increased efficiency, safety, and convenience. As the field of robotics continues to advance, it is crucial to address the challenges and considerations surrounding ethics, legalities, and societal impact to ensure the responsible and beneficial integration of robots into our lives.

## 13.14    Future Perspectives and Emerging Trends

The field of robotics is continuously evolving, and several exciting trends and future perspectives are shaping its trajectory:

Artificial Intelligence and Machine Learning: Integration of artificial intelligence (AI) and machine learning (ML) techniques is revolutionizing robotics. Robots are becoming more capable of learning from data, adapting to new environments, and making intelligent decisions. This trend opens up possibilities for advanced autonomous systems that can perceive, reason, and learn from their experiences.

Humanoid Robotics: Humanoid robots, designed to resemble and interact with humans, are advancing rapidly. With advancements in materials, actuators, and AI, humanoid robots are becoming more dexterous, expressive, and capable of natural interaction. They hold promise in areas such as healthcare, social assistance, and entertainment.

Soft Robotics: Traditional robots are typically made of rigid materials, but soft robotics is an emerging field that focuses on developing robots with soft and flexible structures. Soft robots have inherent compliance and adaptability, enabling them to interact safely with humans, handle delicate objects, and navigate complex environments. This field holds potential in areas such as medical applications and human-robot collaboration.

Swarm Robotics: Swarm robotics involves the coordination of large numbers of simple robots to accomplish complex tasks collectively. Inspired by social insects like ants and bees, swarm robots can exhibit emergent behaviors, self-organization, and fault tolerance. Swarm robotics has applications in areas such as environmental monitoring, search and rescue, and distributed sensing.

Human-Robot Collaboration and Coexistence: The future of robotics lies in seamless collaboration between humans and robots. Collaborative robots or cobots are designed to work alongside humans, sharing workspaces and tasks. Ensuring safe and effective human-robot collaboration is a key area of research, enabling robots to assist, augment, and enhance human capabilities in various domains.

Social and Emotional Interaction: As robots become more integrated into our daily lives, the ability to understand and respond to human emotions becomes crucial. Emotional intelligence and social interaction skills are being incorporated into robotic systems to create more engaging and empathetic experiences. Social robots are being developed to provide companionship, support therapy, and assist individuals in emotional well-being.

Robotics in Extreme Environments: Robots are increasingly being deployed in extreme environments where human presence is challenging or dangerous. From deep-sea exploration to outer space missions, robots are equipped to withstand harsh conditions and perform tasks that would be otherwise impossible for humans. Advancements in ruggedness, autonomy, and sensor technologies enable robots to explore and operate in extreme environments.

Ethical and Responsible Robotics: As robots become more autonomous and intelligent, ethical considerations become more paramount. Developing frameworks and guidelines for responsible robot design, operation, and decision-making is crucial to ensure the ethical use of robotic systems. Addressing concerns such as transparency, accountability, and privacy is vital for building trust and acceptance in society.

In conclusion, the field of robotics is expanding at a rapid pace, with a wide range of applications and exciting future perspectives. Advancements in artificial intelligence, humanoid robotics, soft robotics, swarm robotics, and human-robot collaboration are driving innovation and transforming industries. As robotics technology continues to progress, it is essential to consider ethical, legal, and societal implications to ensure the responsible and beneficial integration of robots into our lives. The future holds immense potential for robotics to enhance our productivity, well-being, and exploration of new frontiers.Nehmzow 2012; Kelly 2013; Bräunl 2003; Bajd et al.

2010; Duffy 1996; Faust 2007; Murray et al. 1994

# Bibliography

Bajd, Tadej et al. (2010). *Robotics*. Vol. 43. Springer Science & Business Media.

Bhaumik, Arkapravo (2018). *From AI to robotics: mobile, social, and sentient robots*. CrC Press.

Bräunl, Thomas (2003). *Embedded robotics*. Springer.

Corke, Peter (2021). *Robotics and Control: Fundamental Algorithms in MATLAB®*. Vol. 141. Springer Nature.

Dahiya, Ravinder S and Maurizio Valle (2013). *Robotic tactile sensing: technologies and system*. Springer.

Davies, Bill (1997). *Practical Robotics*. Werd Technology, Incorporated.

Duffy, Joseph (1996). *Statics and kinematics with applications to robotics*. Cambridge University Press.

Fairhurst, Michael C (1988). *Computer vision for robotic systems: an introduction*. Prentice Hall International (UK) Ltd.

Faust, Russel A (2007). *Robotics in surgery: history, current and future applications*. Nova Publishers.

Jazar, Reza N (2010). *Theory of applied robotics*. Springer.

Katevas, Nikos (2001). *Mobile robotics in healthcare*. Vol. 7. IOS Press.

Kelly, Alonzo (2013). *Mobile robotics: mathematics, models, and methods*. Cambridge University Press.

Liljebäck, Pål et al. (2013). *Snake robots: modelling, mechatronics, and control*. Springer.

Mair, Gordon M (1988). *Industrial robotics*. Prentice Hall.

Mittal, Vibhu O et al. (1998). *Assistive technology and artificial intelligence: applications in robotics, user interfaces and natural language processing*. 1458. Springer Science & Business Media.

Moore, S et al. (2010). "Underwater robotics". In: *Science, Design and Fabrication. Marine Advanced Technology Education Center (MATE), Monterrey CA, USA*.

Murphy, Robin R (2014). *Disaster robotics*. MIT press.

— (2019). *Introduction to AI robotics*. MIT press.

Murray, Richard M et al. (1994). *A mathematical introduction to robotic manipulation*. CRC press.

Nehmzow, Ulrich (2012). *Mobile robotics: a practical introduction*. Springer Science & Business Media.

Niku, Saeed B (2020). *Introduction to robotics: analysis, control, applications*. John Wiley & Sons.

Selig, Jon M (2005). *Geometric fundamentals of robotics*. Vol. 128. Springer.

Siegwart, Roland, Illah Reza Nourbakhsh, and Davide Scaramuzza (2011). *Introduction to autonomous mobile robots*. MIT press.

Stadler, Wolfram (1994). *Analytical robotics and mechatronics*. McGraw-Hill, Inc.

Unbehauen, Heinz (2009). *Control systems, robotics and automation*. Eolss Publishers Company Limited Oxford.

Vertut, Jean and Philippe Coiffet (1986). *Teleoperations and robotics: evolution and development*. Prentice-Hall, Inc.

Vukobratovic, Miomir and Dragan Stokic (2012). *Applied control of manipulation robots: analysis, synthesis and exercises*. Springer Science & Business Media.